Editor
Gisela Lee, M.A.

Managing Editor
Karen Goldfluss, M.S. Ed.

Editor-in-Chief
Sharon Coan, M.S. Ed.

Illustrator
Sue Fullam

Cover Artist
Barb Lorseyedi

Art Director
CJae Froshay

Art Coordinator
Kevin Barnes

Imaging
Rosa C. See

Product Manager
Phil Garcia

Publisher
Mary D. Smith, M.S. Ed.

Author

Mary Rosenberg

Teacher Created Resources, Inc.
6421 Industry Way
Westminster, CA 92683
www.teachercreated.com

ISBN: 978-0-7439-3722-1

Reprinted, 2013
Made in U.S.A.

Table of Contents

Introduction and Materials List

Introduction

The games and activities in *Practice Makes Perfect: Math Games (Grade 2)* focus on important math skills that every second grader needs to learn. Many of the games can be played with only one player or with a partner and use many items commonly found in the home. The games cover some of the concepts listed below.

- logical thinking
- word problems
- addition
- subtraction
- multiplication
- division
- fractions
- estimating
- money
- place value
- time
- measuring

Many of the games do not have just one right answer. This allows children (and parents) of all different abilities and ages to play the games together with enjoyment and success.

Most of the games can be played in multiple ways and can be "custom tailored" to meet the individual child's needs just by changing the numbers being added or subtracted, by changing a manipulative part—spinner, dice, number cards—or by using different numbers on the playing board.

Materials List

- shoebox in which to store game pieces
- plastic sandwich bags in which to store game pieces
- 3" x 5" blank index cards
- counters: beans, bottle caps, paper clips, pennies, pumpkin seeds, sunflower seeds, etc.
- money stamps or stickers
- craft sticks
- glue or hot glue gun
- dice
- playing cards
- 1" graph paper
- 1 cm graph paper
- crayons
- markers
- containers of different sizes
- small items—popcorn kernels, jelly beans, cotton balls, marbles, magnets, etc.
- scratch paper
- inch ruler
- centimeter ruler
- coins (real, plastic, or paper)

Game 1

Estimating Fun

Number of Players: 1 or more

Skill: estimating amounts

Materials

- jars or containers of different sizes
- small items to fill the jars or containers
- scratch paper

Directions

- Fill one of the jars with one of the selected items. (Don't count the items as you fill the jar. Even better—have somebody else fill the jar for you!)
- On a piece of scratch paper, write down your estimate (guess) about how many items are in the jar.
- Empty the jar and sort the items into groups of tens and ones. Compare the actual number to your estimate. Were you close? Were you way off?
- Using another jar that is larger (or smaller) than the first jar, fill the jar with the same items. Figure out a new estimate. Empty the jar and sort the items into groups of tens and ones. Compare your estimate. Were you close? Were you way off?

  Repeat this step with jars or containers of different sizes. The more you practice estimating, the more accurate your estimates will be!

Variation: This activity can be repeated using different items and different sizes of containers.

Game 2

Going on a Scavenger Hunt!

Number of Players: 1 or more

Skill: items to the nearest inch (or centimeter)

Materials

- small items to measure
- 1" graph paper or 1 cm paper (optional)

Object of the Game

- to be the first player to find items that are 1"–12" long (or from 1 cm–30 cm long). You need to find one item for each measurement length.
  Example: 1" = paperclip, 2" = toothpick, 3" = playing card.

Directions

Give each player a piece of graph paper and a pencil or crayon. Have each player look around the house for items to measure. Once the player has found an item to measure, he or she draws a picture of it on the graph paper, returns the item and then continues to hunt for other items to measure.

Game 3

Reading a Graph

Use the graph to answer the questions.

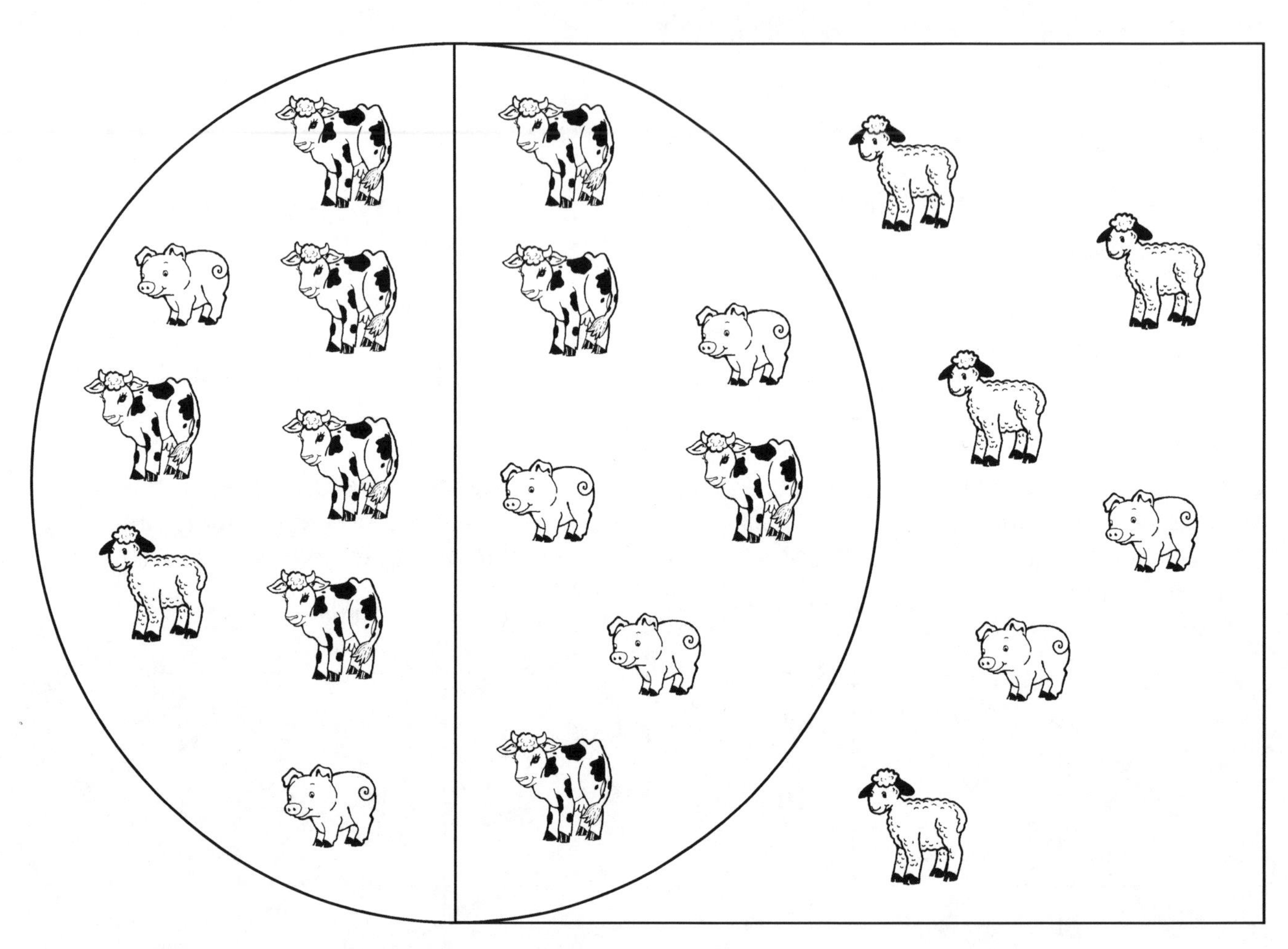

1. Write the number of animals.

 ________  ________ ________

2. How many sheep are in the circle? ___________
3. How many pigs are in the square? ___________
4. How many cows are in the circle but not in the square? ___________
5. How many pigs are in both the circle and the square? ___________
6. How many cows are in the square? ___________
7. Which animal is in the circle and in the square but not in both the circle and the square? ___________

Game 4

Arabic and Roman Numerals

Arabic and Roman Numerals

1 = I	14 = XIV	80 = LXXX
2 = II	15 = XV	90 = XC
3 = III	16 = XVI	100 = C
4 = IV	17 = XVII	150 = CL
5 = V	18 = XVIII	200 = CC
6 = VI	19 = XIX	300 = CCC
7 = VII	20 = XX	400 = CD
8 = VIII	25 = XXV	500 = D
9 = IX	30 = XXX	600 = DC
10 = X	40 = XL	700 = DCC
11 = XI	50 = L	800 = DCCC
12 = XII	60 = LX	900 = CM
13 = XIII	70 = LXX	1,000 = M

Play Roman Numeral Concentration! Take 20 blank 3" x 5" index cards. On 10 of the cards, use Arabic numerals to write the numbers 1–10. On the remaining cards, use Roman numerals to write the numbers 1–10. Shuffle the cards and lay them on the table in a 4 x 5 grid. Taking turns, turn over two cards. If the numerals match, keep both cards and take another turn. If the cards do not match, turn them back over and let the other player take a turn. The player with the most matches wins the game.

Practice writing Roman numerals! Use Roman numerals to write math problems, phone numbers, addresses, birthdays, TV channels, house numbers, etc.

Go on a Roman numeral hunt! Look for Roman numerals inside book covers, at the end of the movie credits, in the newspaper, etc.

Game 5

Secret Message

Use the code to discover the secret message.

VI	II	X	V	III	IX
A	E	I	L	M	N

IV	VII	I	VIII
O	R	S	T

X VIII ' I   VI V V

VII IV III VI IX

VIII IV   III II !

Game 6

What's in a Name?

Number of Players: 1 or more

Skills

- counting coins
- adding coins

Materials

- scratch paper
- real, plastic, or paper coins
- value chart

Object of the Game

- to find the value of different coins by names and words

Directions

Have the player write his or her name on a piece of paper. Using the chart below, have the playe ure out the "value" of his or her name.

Variations

- Try to find names and words worth a specific value. Examples: 50¢ or 25¢.
- Make your own value chart using 1" graph paper and paper or sticker coins or coin stamps. Compare the same words using the old value chart and the new one. What happened to the value of each word? Did the value change? Why or why not?
- Make the activity easier by making a value chart using only pennies and nickels.
- Make the activity more challenging by making your own value chart using different amounts of money. Example: A = 6¢, B = 9¢, C = 3¢

Value Chart

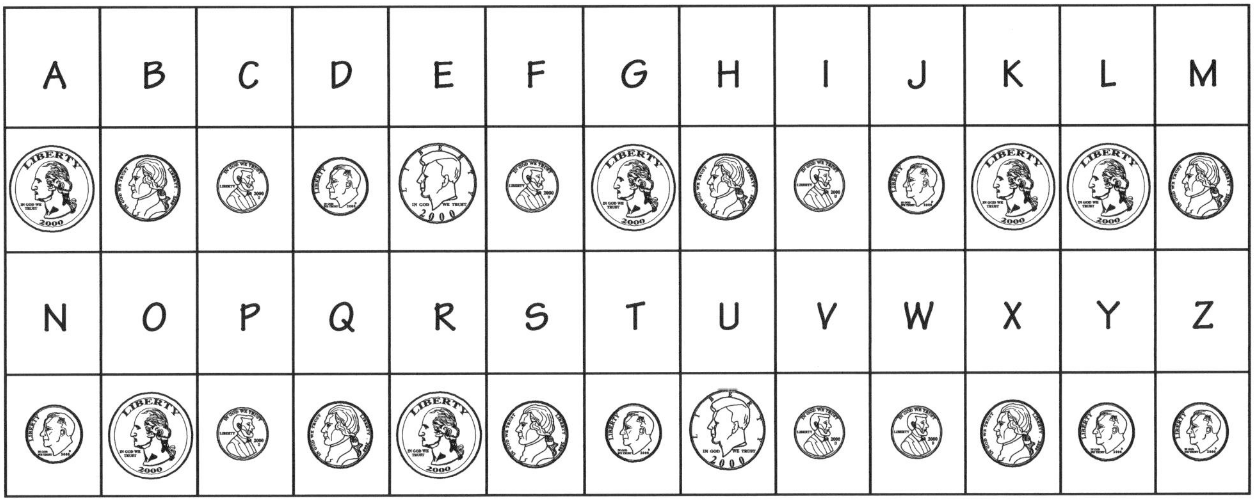

A	B	C	D	E	F	G	H	I	J	K	L	M
N	O	P	Q	R	S	T	U	V	W	X	Y	Z

Game 7

Money Madness

Use real, plastic, or paper coins to solve each money riddle.

1. La Tonya has 8 pennies, 4 nickels, 5 dimes, and 2 quarters. How much money does La Tonya have? La Tonya has ____________.	2. Gregory has 2 pennies, 6 nickels, 6 dimes, and 0 quarters. How much money does Gregory have? Gregory has ____________.
3. Molly has 5 pennies, 7 nickels, 0 dimes, and 1 quarter. How much money does Molly have? Molly has ______________.	4. Matilda has 0 pennies, 5 nickels, 2 dimes, and 2 quarters. How much money does Matilda have? Matilda has ______________.
5. Miles has 4 pennies, 0 nickels, 10 dimes, and 1 quarter. How much money does Miles have? Miles has ______________.	6. Ross has 10 pennies, 3 nickels, 10 dimes, and 2 quarters. How much money does Ross have? Ross has ______________.

Game 8

Even More Money Madness!

Use real, plastic, or paper coins to solve each money riddle.

1. Michelle has 5 coins in her pocket. Together the coins equal 47¢. What coins does Michelle have in her pocket?

2. Kevin has 3 coins in his pocket. The coins are all of the same value. The coins equal 75¢. What coins does Kevin have in his pocket?

3. Rodney has 4 coins in his pocket. The coins are all of different values. The coins equal 41¢. What coins does Rodney have in his pocket?

4. Nancy has 2 coins in her pocket. The coins are of the same value and equal $1.00. What coins does Nancy have in her pocket?

5. Wanda has 6 coins of three different values. Together the coins equal 80¢. What coins does Wanda have in her pocket?

6. Brandon has 7 coins in his pocket. Together the coins equal 51¢. What coins does Brandon have in his pocket?

Game 9

Make a Tie

Number of Players: 1 or more

Skills

- counting money
- subtracting money

Price List	
stripes = 3 for 10¢	tie tack = 8¢
large dots = 2 for 5¢	tie pin = 7¢
small dots = 2 for 3¢	tie clip = 4¢
pattern = 15¢	background color = 20¢
picture = 15¢	

Materials

- scratch paper
- crayons or markers
- 6-sided die
- price list
- *optional:* real, plastic, or paper coins
- *optional:* calculator

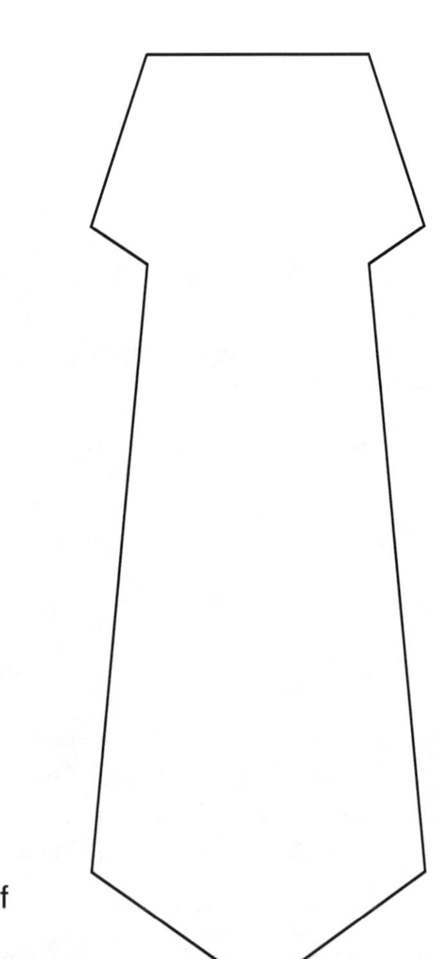

Object of the Game

- to decorate a tie and stay within a budget

Directions

- Roll a die to determine what each student's budget will be.

6-Sided Die		
1 = 25¢	3 = 75¢	5 = $1.25
2 = 50¢	4 = $1.00	6 = $1.50

- Draw a tie on a piece of scratch paper.
- Use the price list to figure out what can be bought with a specific amount of money. Add the items to the tie—without going over budget!

Variations

- This same activity can be done using pennies instead of quarters. **Example:** 1 = 1¢, 2 = 2¢, 3 = 3¢, 4 = 4¢, 5 = 5¢, and 6 = 6¢. Change the price list to reflect the different amounts of money.
- Instead of using a tie, use pet supplies and a picture of a puppy or kitten. **Suggested items to use:** collar, brush, food, bed or basket, water bottle, toys, etc. Just develop a price list to fit the new item(s).

Game 10

Money Toss

Number of Players: 1 or more

Skills

- using probability
- making a graph to show information (see the sample below)
- comparing estimate to actual outcome

Materials

- 1" graph paper
- 10 nickels
- crayons or markers
- scratch paper (optional)

Directions

- Make a guess as to the number of heads that will be showing after you toss 10 nickels. Write the guess on the graph paper or a piece of scratch paper.
- Toss 10 nickels on the carpet or on a table. Count only the nickels that are heads up. Add together their total value. Color the matching square on the graph. (Four heads-up nickels, color the 20¢ square.) Repeat this step nine more times. Compare the actual number of times the nickels landed heads up to your estimate. Write a sentence reporting your findings.

Variation

Repeat this same experiment using pennies, dimes, and quarters. What do you notice about the results?

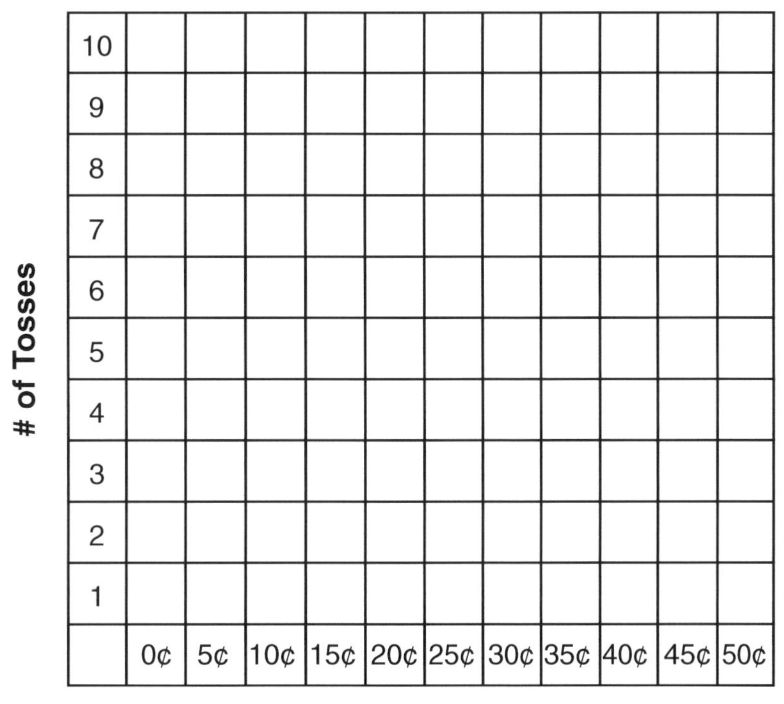

Game 11

They're Mad About Math

Find and color each word in the list below.

A	D	A	P	R	O	D	U	C	T	Q	F
D	I	F	F	E	R	E	N	C	E	U	A
D	G	B	M	O	C	Q	U	S	V	O	C
D	I	V	I	D	E	U	M	U	E	T	T
D	T	E	N	D	F	A	B	B	N	I	O
J	S	L	U	I	O	L	E	T	G	E	R
K	S	M	S	N	N	P	R	R	S	N	T
H	U	N	D	R	E	D	S	A	V	T	T
W	M	X	Y	P	S	A	B	C	C	D	E
T	M	P	O	L	N	M	L	T	E	Y	N
S	R	Q	M	U	L	T	I	P	L	Y	S
T	H	O	U	S	A	N	D	K	J	I	H

1. add
2. difference
3. digits
4. divide
5. equal
6. even
7. factor
8. hundreds
9. minus
10. multiply
11. numbers
12. odd
13. ones
14. plus
15. product
16. quotient
17. subtract
18. sum
19. tens
20. thousand

Game 12

Picking Beans

Number of Players: 1 or more

Skill: place value

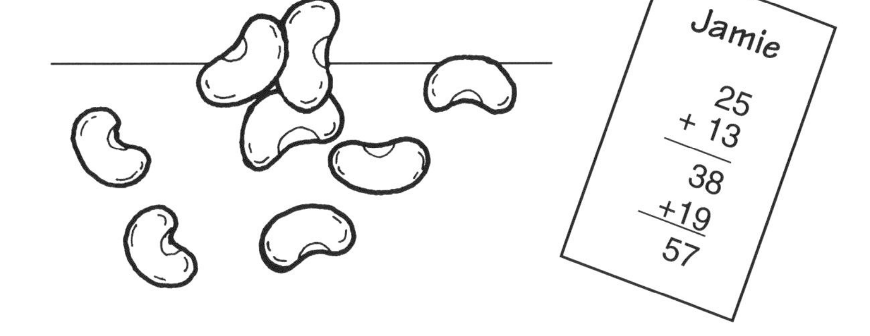

Materials

- small bowl
- bag of beans
- scratch paper

Object of the Game

- to reach 100 (or some other specified number) first

Directions

Pour the beans into the bowl. Reaching in with one hand, the player grabs as many beans as possible. The player then sorts the beans into groups of ten and the remaining beans are the ones. The player writes the number on a piece of scratch paper. Repeat these same steps and add the two numbers together. The first person to reach 100 wins the game!

"Bean Sticks"

"Bean sticks" can be used to practice regrouping skills. You will need some beans, craft sticks, and white glue or a glue stick. Glue 10 beans onto each bean stick. Each bean stick is worth "1 ten." Leave a handful of beans to act as the "ones." Use the bean sticks to solve addition and subtraction problems where regrouping is needed.

$$\begin{array}{r} 28 \\ +\ 15 \\ \hline \end{array}$$

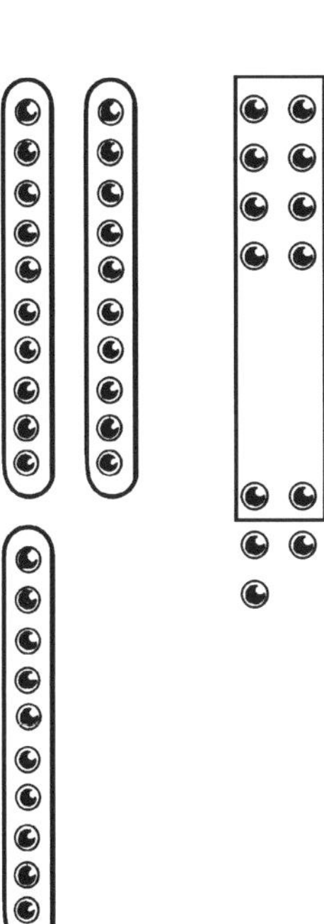

Exchange the 10 "ones" for a bean stick. Write the number of tens and ones.

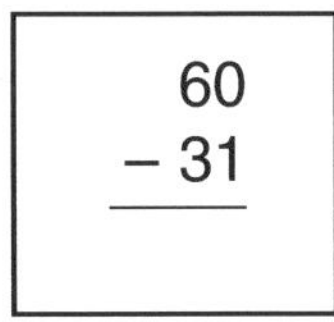

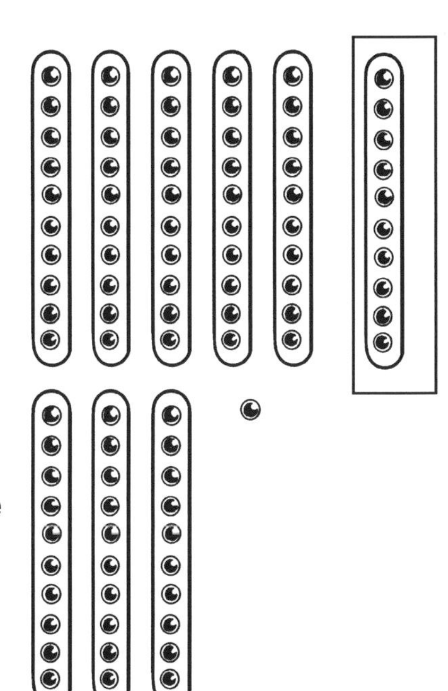

Exchange one of the "tens" for 10 "ones." Subtract the number of ones and then the tens. Write the answer.

Game 13

Make a Hundred

Number of Players: 1 or 2

Skill: adding numbers to 100

Materials

- playing board
- crayons or markers (or counters can be used)

Object of the Game

- to add the numbers in each square to make 100. (The squares must be touching on at least one side. The squares may be colored or can be covered with a counter.)

Directions

- If playing with a partner, the winner is the player who makes the most hundreds.
- If playing the game alone, the player uses different colors of crayons to color each set of numbers that make 100.
- If playing with a partner, each player uses a different color crayon, marker, or counter.

(10)	(5)	50	10	5	50
(25)	(50)	5	25	10	10
(5)	(10)	(5)	5	5	50
(50)	(5)	(25)	25	5	5
(5)	(10)	50	50	25	50
25	50	5	10	50	10
25	25	5	25	50	5
5	50	10	25	10	25

Variations

- Play the game adding numbers to 50.
- On 20 blank 3" x 5" index cards, write the following number words: five, ten, fifteen, twenty, twenty-five, thirty, thirty-five, forty, forty-five, fifty. Playing alone or with a partner, the player turns over a card and colors the corresponding number (or numbers).
- Play the game with the winner being the player who has colored (or covered) five squares in a row—either horizontally, vertically, or diagonally.
- Use 1" graph paper to make your own playing board using money amounts—5¢, 10¢, 25¢—coin stamps or stickers, or larger numbers—100, 200, 300, etc.

Game 13 *(cont.)*

Make a Hundred Playing Board

10	5	50	10	5	50
25	50	5	25	10	10
5	10	5	5	5	50
50	5	25	25	5	5
5	10	50	50	25	50
25	50	5	10	50	10
25	25	5	25	50	5
5	50	10	25	10	25

Game 14

The 25-50 Game

Number of Players: 2 or more

Skill: skip counting

Materials

- game board
- 20 different counters for each player (beans, pennies, sunflower seeds, pumpkin seeds, paper clips, bottle caps, etc.)
- 40 blank 3" x 5" index cards with tally marks, coin stamps, coin stickers, Roman numerals, or math problems (See sample cards below.)

Object of the Game

- to be the first player to have 5 counters in a row

Directions

Taking turns, each player turns over a card and places a counter on the corresponding space.

Variations

- Have each player make a specific number or amount of money on each turn—for example 100 or $1.00—make sure the spaces are touching on at least one side!
- Make extra cards showing different amounts, such as 75 (or 75¢), 125 (or $1.25).

Coins

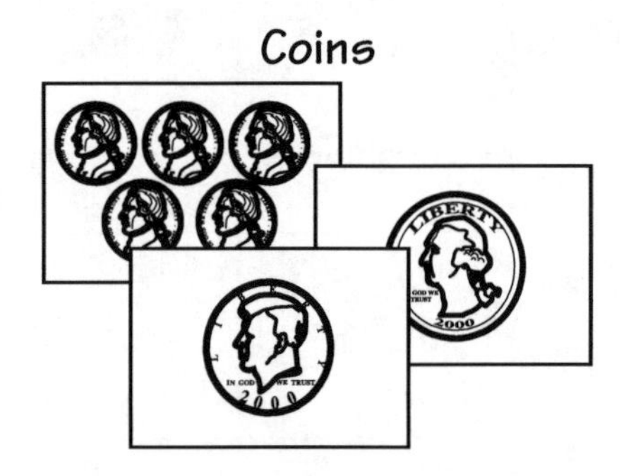

using coin stamps or stickers

Math Problems

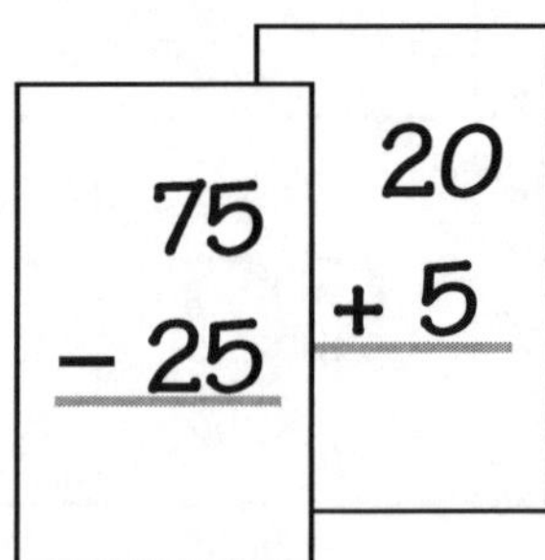

Tally Marks

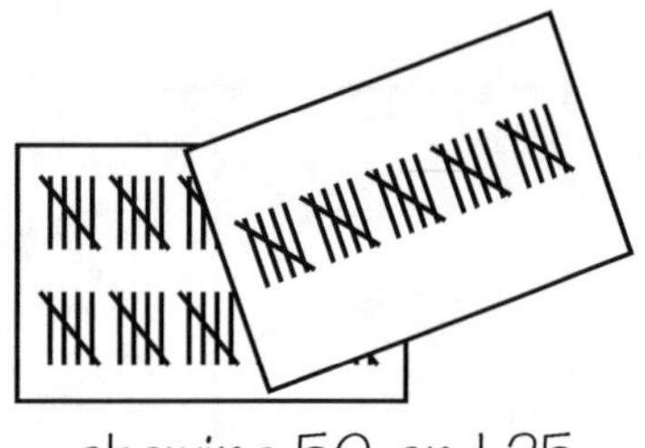

showing 50 and 25

Roman Numerals

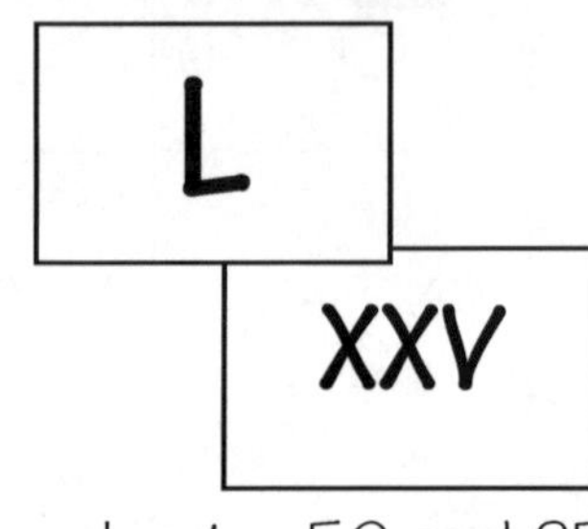

showing 50 and 25

Game 14 *(cont.)*

The 25-50 Game Board

50	25	50	25	50	25
25	50	25	50	25	50
25	50	25	50	25	50
25	50	50	25	50	25
50	25	50	25	25	50
25	50	25	25	50	50

Game 15

Hundreds of Riddles

Use the hundreds board (page 21) to solve each riddle.

Hundreds Riddle #1

- The number is smaller than 73 – 15.
- The number is larger than 45 – 4.
- The difference between the two digits in the number is 5.
- The sum of the two digits in the number is 5.

What is the mystery number? _______

Hundreds Riddle #2

- The number is smaller than 71 + 3.
- The number is larger than 19 + 41.
- It is an odd number.
- The sum of the two digits in the number is 10.
- The difference between the two digits in the number is 4.

What is the mystery number? _______

Hundreds Riddle #3

- The number is larger than 10 x 4.
- The number is smaller than 60 + 10.
- The number is divisible by 2.
- When the two digits are multiplied together the product is 8.

What is the mystery number? _______

Hundreds Riddle #4

- The number is a multiple of 2.
- The number is a multiple of 5.
- The number is larger than 10 x 3.
- The number is smaller than 5 x 10.

What is the mystery number? _______

Game 15 *(cont.)*

Hundreds Riddles Board

Solve each riddle by reading each clue and using markers (beans, dimes, etc.) to cover the numbers. To make your own hundreds riddles, use 1" graph paper and write a number in each box. To make the riddles easier, use fewer numbers—like 20 to 40—or use only addition or subtraction problems in the clues.

1	2	3	4	5	6	7	8	9	10
11	12	13	14	15	16	17	18	19	20
21	22	23	24	25	26	27	28	29	30
31	32	33	34	35	36	37	38	39	40
41	42	43	44	45	46	47	48	49	50
51	52	53	54	55	56	57	58	59	60
61	62	63	64	65	66	67	68	69	70
71	72	73	74	75	76	77	78	79	80
81	82	83	84	85	86	87	88	89	90
91	92	93	94	95	96	97	98	99	100

Game 16

Magic Nines

Complete the square using the numbers: 0, 10, 20, 30, 40, 50, 60, 70, 80, 90, or 100. Each row and column must add up to the specific number listed below. (*Hint:* Write the numbers on sticky notes so they can be easily moved around!)

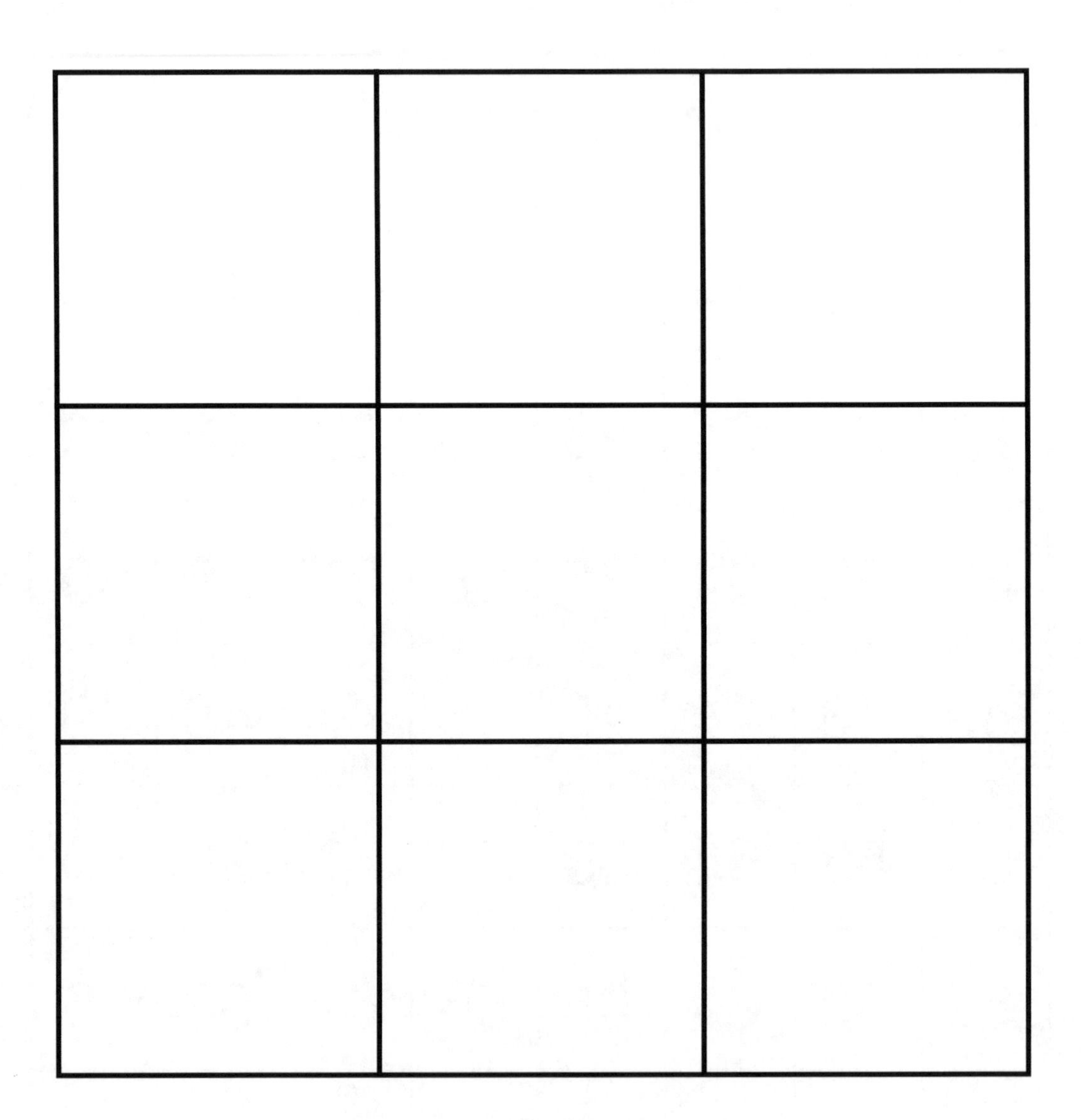

1. The Magic Number is 120.
2. The Magic Number is 150.
3. The Magic Number is 180.

Game 17

Classroom Logic

Read the clues to discover each student's classroom number.
Mark an "O" for yes.
Mark an "X" for no.

	2	11	26	37	58
Judy					
Abe					
Marilyn					
Bud					
Kitty					

1. Judy is in an even-numbered classroom.
2. Abe is in an odd-numbered classroom.
3. Marilyn is in a higher-numbered classroom than Kitty.
4. Bud is in Room 37.
5. Judy is in a higher-numbered classroom than both Kitty and Marilyn.

Write the classroom number next to each student's name.

Judy ______

Abe ______

Marilyn ______

Bud ______

Kitty ______

Game 18

Knock! Knock!

Read each clue to figure out who lives in each house.

1. Andrew does not live in the houses with the highest or the lowest numbers.
2. Jessica lives in house 425.
3. Peter lives in a house with an even number.
4. Brian lives in a house with a number divisible by 10.
5. Sarah lives in the house with a number 100 lower than Jessica's.
6. Andrew lives in a house with a larger number than Peter's and a smaller number than Brian's.

Draw a line matching each person to his or her house.

Andrew **Brian** **Jessica** **Peter** **Sarah**

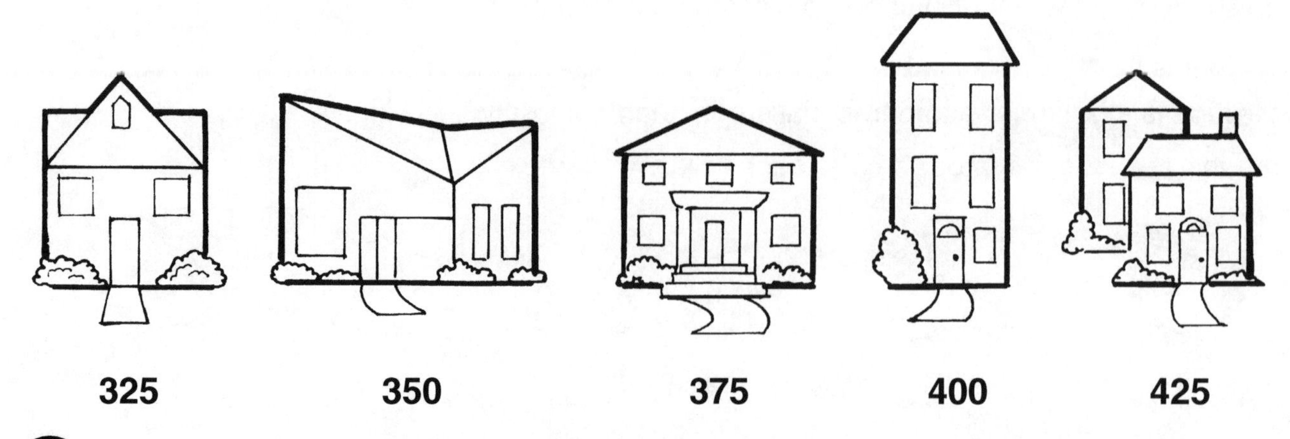

Game 19

My Favorite Number

Use the clues to find each person's favorite number.

0	10	5	13
9	4	1	7
6	12	14	3
2	15	8	11

Mystery Number #1

- Anthony's favorite number is an even number.
- The number is greater than 10 and less than 14. What is his favorite number?

____

Mystery Number #2

- Reba's favorite number is an odd number with two digits.
- When the two digits are added together the sum is 4. What is her favorite number? ___________

Mystery Number #3

- Lucy's favorite number is a multiple of 5.
- The number is also a multiple of 2 and 10. What is her favorite number?

____

Mystery Number #4

- Joey's number is divisible by 2, 3, and 4.
- What is Joey's favorite number?

____

Game 20

Does Anybody Know the Time?

Rewrite each time in standard form.

1. A quarter til 9: ________________
2. Half past 3:________________
3. A quarter past 4:________________

Find the time for each event.

1. **Movie Times**

Paperboy Always Delivers	8 : 0 0
Scarier than Scary	___:___ ___
Home But Not Alone	___:___ ___

- The *Paperboy Always Delivers* starts 1/2 hour earlier than *Scarier than Scary*.
- *Home But Not Alone* starts 1/2 later than *Scarier than Scary*.

2. **Amusement Park Rides**

Roller Coaster	9 : 4 5
Elevator Ride	___ ___:___ ___
Going Around	___ ___:___ ___

- The Elevator Ride starts 15 minutes after the Roller Coaster ride.
- Going Around starts 30 minutes after the Roller Coaster ride.

3. **Water Park**

red boards	3 : 1 5
yellow boards	___:___ ___
blue boards	___:___ ___

- The yellow boards need to be returned 15 minutes after the red boards.
- The blue boards need to be returned 15 minutes before the red boards.

4. **Story Theater**

Paper Bag Theater	___:___ ___
Puppet Theater	___:___ ___
Sock Puppet Theater	4 : 3 0

- Sock Puppet Theater begins 15 minutes after Puppet Theater.
- Paper Bag Theater begins 15 minutes before Puppet Theater.

Game 21

2-by-2 Game

Number of Players: 2–4

Skill: multiplying by 2's

Materials

- game board
- marker for each player
- make several sets of the game cards

Object of the Game

- to be the first player to reach the ark

Directions

- Each player places a marker in the Start box.
- Taking turns each player turns over a card, solves the problem, and moves to the nearest space with the matching answer. If another player lands in the same space, the first player has to go back to Start.

Variations

- This game can be played in a variety of ways and with a variety of math problems. Some sample problems are listed below.

Multiplying by 2's		Addition Doubles to 20		
2 x 0	2 x 6	0 + 0	4 + 4	8 + 8
2 x 1	2 x 7	1 + 1	5 + 5	9 + 9
2 x 2	2 x 8	2 + 2	6 + 6	10 + 10
2 x 3	2 x 9	3 + 3	7 + 7	
2 x 4	2 x 10			

- Division problems can also be used.
- To make the game more challenging and/or to review math facts, make game cards with "mixed-practice" problems—addition, subtraction, multiplication, and division problems. Play the game as outlined above.
- Take 30 blank 3" x 5" index cards. Make several sets of the following number words: zero, two, four, six, eight, ten, twelve, fourteen, sixteen, eighteen, twenty. Play the game as outlined above.

Game 21 (*cont.*)

2-by-2 Game Board

Start →

14

4

6

18

0

12

2

8

20

0

10

16

18

14

0

2

20

8

16

10

12

4

6

2-by-2

Game 22

Target Practice

Find the factors that equal each product.

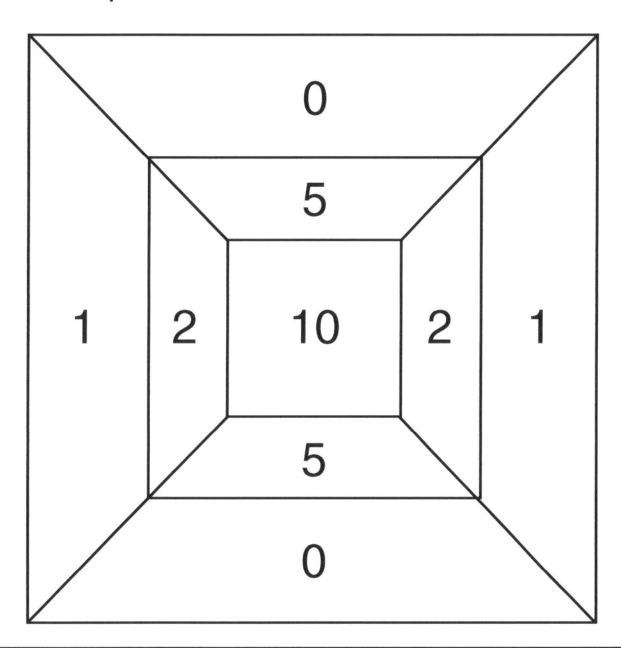

1. In four turns, Amelia scored a total of 20 points. Which numbers did she hit?

   Amelia hit ____, ____, ____, and ____.

2. In three turns Scout scored a total of 75 points. Which numbers did Scout hit?

   Scout hit _____, _____, and _____.

3. Rachel scored a total of 100 points in two turns. Which numbers did Rachel hit?

   Rachel hit _____ and _____.

4. Ellis scored 8 points in 3 turns. Which numbers did Ellis hit?

   Ellis hit _____, _____, and _____.

5. Ross scored a total of 20 points in three turns. Which numbers did Ross hit?

   Ross hit _____, _____, and _____.

6. Mona had the highest score. She scored 1,000 points in three turns. Which numbers did Mona hit?

   Mona hit _____, _____, and _____.

Game 23

The Fives Have It

Number of Players: 2

Skill: multiplying by 5's

Materials

- game board
- crayons or markers
- 6-sided die

Object of the Game

- to use all of the squares on the board

Die

Multiply the number showing on the die by 5.

x	1	2	3	4	5	6
5						

Directions

- The player rolls the die and multiplies the number showing on the die by 5. The player then colors in 1, 2, 3, 4, or more squares—that are touching on at least one side—that would match the product. Example: The die shows a 4. 4 x 5 = 20. The player could color in a 5, 10, and another 5.
- If playing with two players using one board:

  Each player would need one crayon or marker. Taking turns, each player rolls the die and colors in the squares that would equal the product. The player with the most squares colored wins the game!
- If playing with two players using two boards:

  Taking turns, each player rolls the die and colors in the squares on his or her own board that would equal the product. The player who colors in his or her board first wins the game!

Variations

- Make your own playing board using 1" graph paper and numbers of your choice. If you need to work on multiplication skills, multiply the number showing on the dice by 2, 3, 4, or whatever number you need to work on.
- If you need to work on addition or subtraction skills, write the sums (or differences) that you need to practice on the graph paper. Select the number that you need to practice—for example 12. Roll the die and add 12 to it or roll the dice and subtract the number showing from 12.
- Use 6 or more nickels to play the game. Toss the die on the table and multiply the number of "heads" showing by the number 5. Color the square(s) to equal the product.

Game 23 (cont.)

The Fives Have It Game Board

5	10	30	20	5	15	10	25	5	30
15	5	20	30	15	10	30	10	5	20
10	30	25	30	5	20	0	10	5	5
15	10	10	5	10	15	10	25	30	5
30	25	5	20	20	10	10	10	5	25
5	10	15	20	25	25	30	15	5	10
10	15	10	5	15	5	25	20	15	20
30	5	20	10	5	20	10	5	30	10
10	10	5	5	25	10	15	10	25	25
10	5	20	25	15	5	30	5	5	15

Game 24

It's All in the Cards

Number of Players: 1 or more

Skill: practicing addition, subtraction, multiplication, and division skills

Materials

- playing cards with the face cards removed
- ten 3" x 5" index cards cut in half. Make 5 sets of cards showing the following operations: +, –, x, and ÷ .
- scratch paper

Object of the Game

- to practice basic math skills

Directions

- Shuffle both sets of cards (order of operation cards and face cards) and place in separate stacks face-down on the table. Turn over the top two playing cards and the top operation card. Write the problem that is showing.

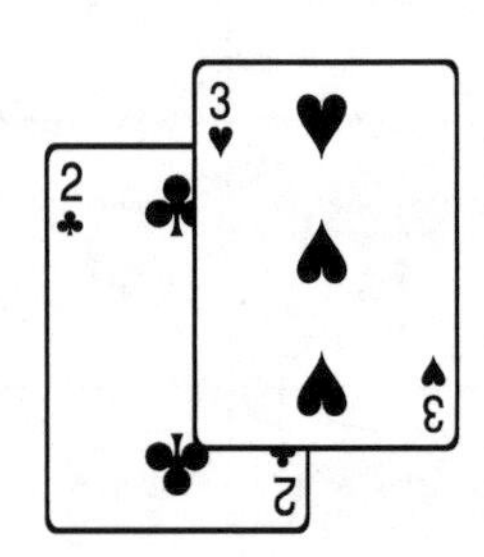

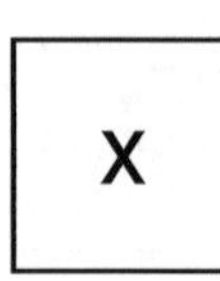

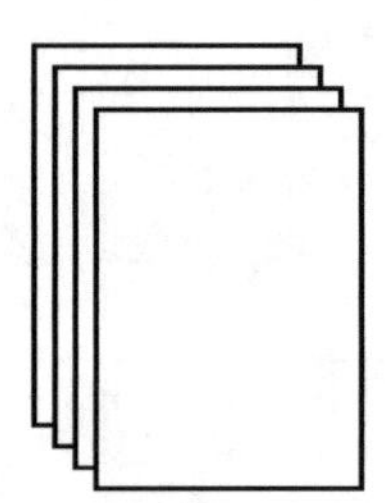

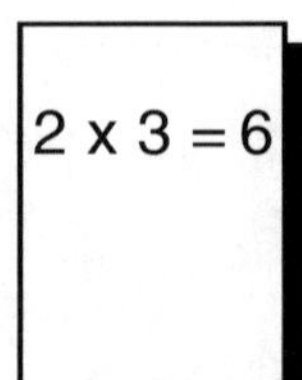

Variations

- To make the game easier, use only one or two operation cards.
- To make the game more challenging, use 4 (or more) playing cards and 3 (or more) operation cards.

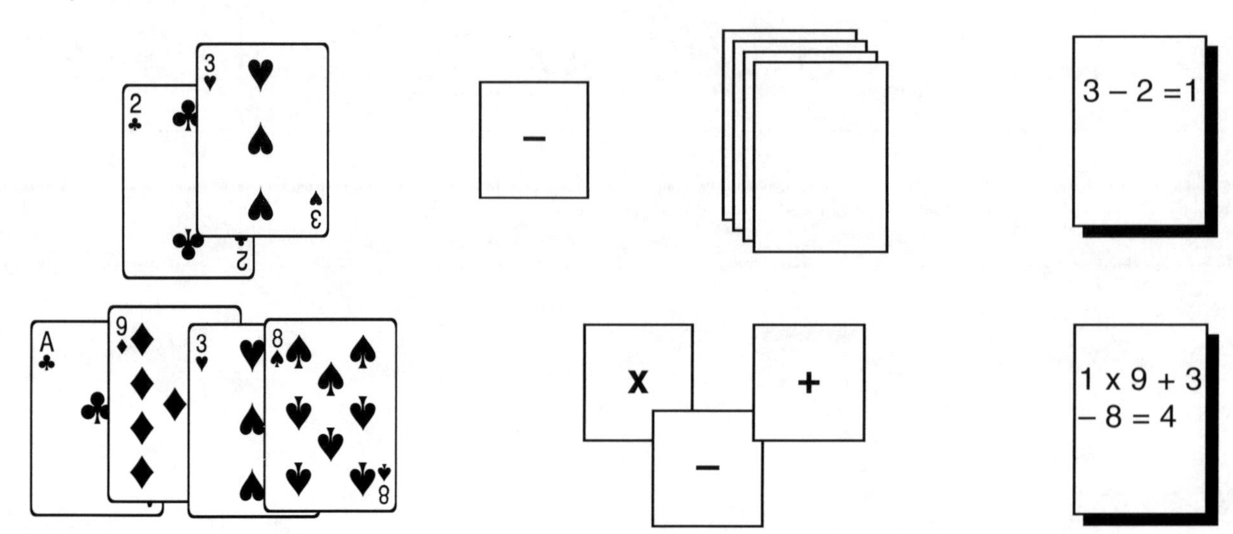

Game 25

Times Ten

Number of Players: 1 or more

Skills

- multiplying by 10
- adding to 1,000
- subtracting from 1,000

Materials

- playing cards with the face cards removed
- scratch paper

Object of the Game

- if adding—to be the first one to reach 1,000
- if subtracting—to be the first one to reach 0

Directions

- Shuffle the cards and place in a stack face-down on the table. Turn over the first card, multiply it by 10, write the product on the scratch paper. Repeat these steps, writing the product on the paper, and adding the two numbers. Continue in this manner until 1,000 is reached.
- If playing with two or more players: The game is played as described above, but the players take turns.

Variations

- Play the game as described above but subtract the product from 1,000. Repeat the steps until is reached.
- For a more challenging game, multiply the playing cards by 100.
- For an easier game, multiply the number by 2.
- To practice addition skills, select a number—20 for example—and add the number showing on the card to it.
- To practice subtraction skills, select a number—18 for example—and subtract the number showing on the card from it.

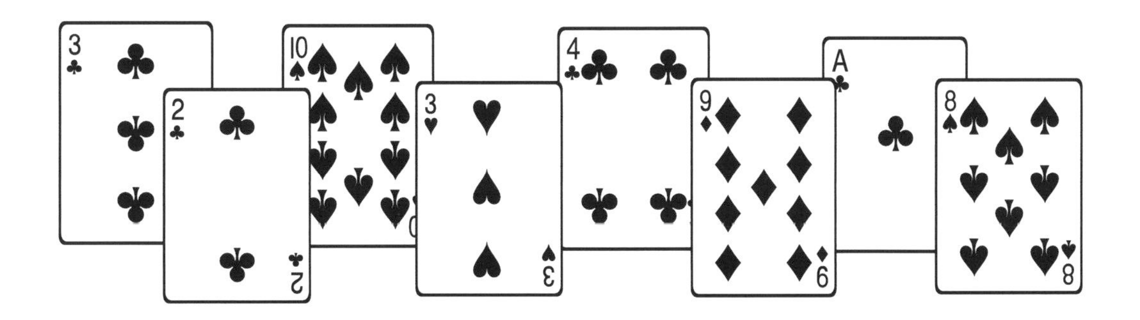

Game 26

What's the Score?

Use division to figure out how many points each person scored in each game.

Final Scores

Volleyball	20 points	Hockey	15 points
Bowling	60 points	Soccer	10 points
Basketball	50 points	Football	12 points

1. Sylvia scored 1/5 of the points in basketball. How many points did Sylvia earn?

   Sylvia earned __________ points.

2. Johnny scored 3/10 of the points in soccer. How many points did Johnny earn?

   Johnny earned __________ points.

3. Greg scored 1/2 of the points in volleyball. How many points did Greg earn?

   Greg earned __________ points.

4. Bonnie scored 1/10 of the points in bowling. How many points did Bonnie earn?

   Bonnie earned __________ points.

5. Annie scored 1/2 of the points in football. How many points did Annie earn?

   Annie earned __________ points.

6. Robbie scored 1/5 of the points in hockey. How many points did Robbie earn?

   Robbie earned __________ points.

Game 27

Find the Number

Color the number of items.

1. Color 3/4 of the 16 shoes. 3/4 of 16 = ________ shoes

2. Color 2/5 of the 15 candies. 2/5 of 15 = ________ candies

3. Color 3/10 of the 10 ice cream cones. 3/10 of 10 = ________ cones

4. Color 1/2 of the 20 skateboards. 1/2 of 20 = ________ skateboards

Game 28

It's All in a Foot

Write the number of inches. Use the >, <, and = symbols to compare the two amounts.

1. 3/4 of a foot = ________ inches   1/2 of a foot = ________ inches

   3/4 of a foot ◯ 1/2 of a foot

2. 1/2 of a foot = ________ inches   3/6 of a foot = ________ inches

   1/2 of a foot ◯ 3/6 of a foot

3. 5/6 of a foot = ________ inches   1/4 of a foot = ________ inches

   5/6 of a foot 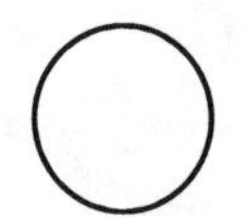 1/4 of a foot.

4. 3/4 of a foot = ________ inches   2/3 of a foot = ________ inches

   3/4 of a foot 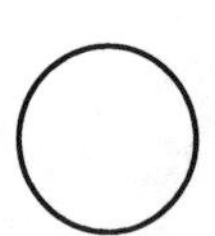 2/3 of a foot

5. 1/3 of a foot = ________ inches   1/6 of a foot = ________ inches

   1/3 of a foot 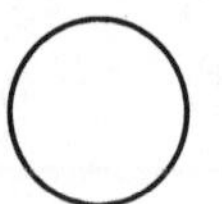 1/6 of a foot

6. 1/4 of a foot = ________ inches   1/3 of a foot = ________ inches

   1/4 of a foot 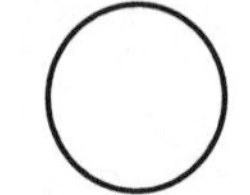 1/3 of a foot

Game 29

What's the Rule?

Finish the pattern and write the rule. Make your own patterns for a family member or friend to solve for the last three boxes.

1.

1	4
3	6
7	
2	

Rule:______________

____

2.

2	4
8	16
9	
1	

Rule:______________

____

3.

25	5
15	3
10	
20	

Rule:______________

____

4.

10	100
3	30
2	
5	

Rule:______________

____

5.

15	12
9	6
3	
7	

Rule:______________

____

6.

6	16
4	14
3	
11	

Rule:______________

____

7.

Rule:______________

____

8.

Rule:______________

____

9.

Rule:______________

____

Game 30

Brain Busters

Write the equation using the correct math symbol (+, –, x, ÷) to solve each problem.

1. Number of legs on 3 dogs.	2. Number of eyes on 1 person.	3. A dozen eggs placed equally in 3 bowls.
4. Number of fingers on 3 hands.	5. How many eyes on five 3-eyed aliens?	6. Number of nickels it takes to make one quarter.
7. Number of cookies in 1/2 a dozen.	8. There are 10 points. How many stars?	9. How many dimes are in one dollar?
10. Number of legs on 5 spiders.	11. Number of legs on 5 insects.	12. Half of one dollar.

Game 31

What's My Sign?

Add the correct operational symbol (+, –, x, ÷) to complete each problem. Each symbol can only be used one time with each set of numbers.

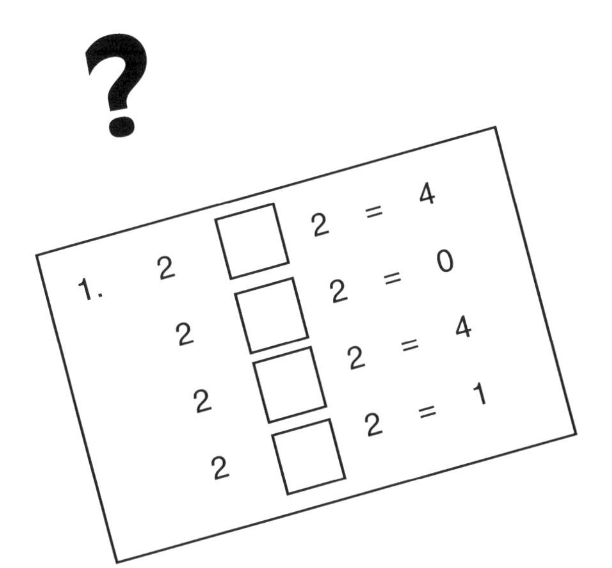

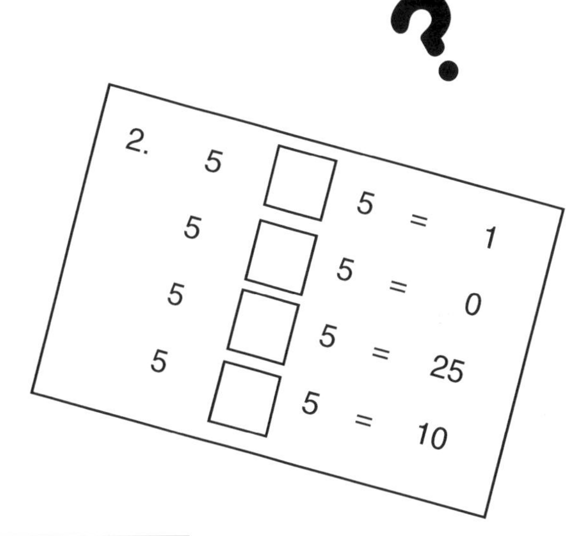

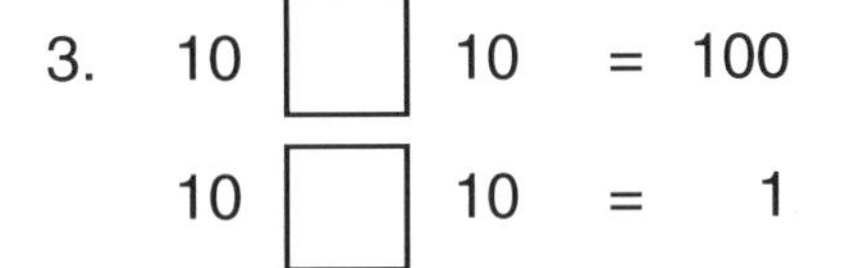

3. 10 ☐ 10 = 100
10 ☐ 10 = 1
10 ☐ 10 = 0
10 ☐ 10 = 20

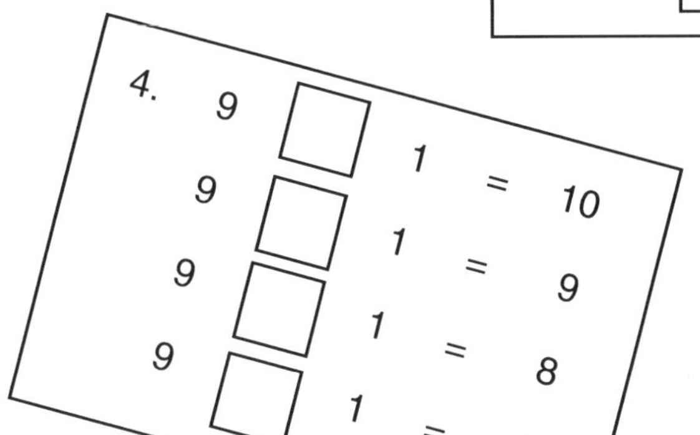

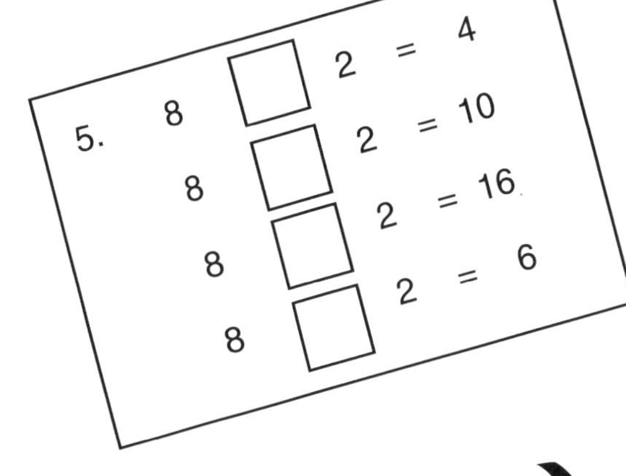

Game 32

Names and Numbers

Read each clue. Write the number word in the crossword puzzle.

Down

1. legs on an octopus
2. a pair
3. wheels on a tricycle
4. 6 + 5
5. 100,000 x 0
6. number of fingers on one hand
9. 3 sets of 3 balls
11. legs on a bug

1 2 3 4 5 6 7 8 9 10 11

Number Words

0 = zero
1 = one
2 = two
3 = three
4 = four
5 = five
6 = six
7 = seven
8 = eight
9 = nine
10 = ten
11 = eleven
12 = twelve

Across

6. 2 x 2
7. XII
8. a single puppy
10. 10 x 1
11. 13 – 6

Game 33

What Did You Say?

One piece of information is missing from each word problem. Write the missing piece of information on the line.

1. Jewel made 15 cookies. Her dog, Ruby, ate some of them. How many cookies does Jewel have left?

   Missing information: ______________________________

   ____

2. Bart bought 3 books of stamps. Each book has some stamps. How many stamps did Bart buy in all?

   Missing information: ______________________________

   ____

3. Martha made a dozen cookies. Jerilyn made some cookies too. How many cookies were made in all?

   Missing information: ______________________________

   ____

4. Larry planted 100 flowers in several rows. How many flowers did Larry plant?

   Missing information: ______________________________

   ____

5. Doug bought a tie. He paid with a $10 bill. How much change was Doug given?

   Missing information: ______________________________

   ____

Game 34

Time for Trivia

Time how long it takes for you to answer the questions. (Write your answers on a separate sheet of paper.) Have a friend or family member answer the same questions. Who had the fastest time?

1. How many days in one week? ____________
2. How many thumbs do you have? ____________
3. Name a shape. ____________
4. What is the value of a penny? ____________
5. What is the day after tomorrow? ____________
6. How many inches in one foot? ____________
7. How many feet in one yard? ____________
8. Which is worth more: 10 dimes or 5 quarters? ____________
9. How long should you boil a 3-minute egg? ____________
10. How much is a half-dollar piece worth? ____________
11. What is the perimeter of a square with sides of 3 feet long? ____________
12. How many seconds in one minute? ____________
13. How many minutes in one hour? ____________
14. How many eyes does a three-eyed monster have? ____________
15. 100,000 x 0 = ? ____________
16. How many months in one year? ____________
17. How many months have fewer than 31 days? ____________
18. How many months have more than 30 days? ____________
19. How many pennies in one dollar? ____________
20. What number is missing? 5, 10, ____, 20, 25, 30

Game 35

Fraction War

Number of Players: 2–4

Skill: comparing fractions

Materials

- fraction equivalent bars
- fraction flashcards from Fraction Wizards or make a new set of cards. To make a new set of cards, cut 39 blank 3" x 5" index cards in half. Write the following fractions on individual cards:

1/1, 1/2, 2/2 1/3, 2/3, 3/3, 1/4, 2/4, 3/4, 4/4, 1/5, 2/5, 3/5, 4/5, 5/5, 1/6, 2/6, 3/6, 4/6, 5/6, 6/6, 1/7, 2/7, 3/7, 4/7, 5/7, 6/7, 7/7, 1/8, 2/8, 3/8, 4/8, 5/8, 6/8, 7/8, 8/8, 1/9, 2/9, 3/9, 4/9, 5/9, 6/9, 7/9, 8/9, 9/9, 1/10 2/10, 3/10, 4/10, 5/10, 6/10, 7/10, 8/10, 9/10, 10/10, 1/11, 2/11, 3/11, 4/11, 5/11, 6/11, 7/11, 8/11, 9/11, 10/11, 11/11, 1/12, 2/12, 3/12, 4/12, 5/12, 6/12, 7/12, 8/12, 9/12, 10/12, 11/12, 12/12

Object of the Game

- to collect all of the cards

Directions

- Shuffle the cards and divide them evenly among all of the players. At the same time, all players turn over the top card. The person with the largest fraction wins the round and keeps all cards from that round.
- If two or more players have the same fraction, those players turn over the next card and compare fractions. The person with the largest fraction wins and gets to take all cards.
- Once a player has used all of the cards he or she was originally given, the player shuffles the cards he or she has collected and this becomes a "fresh" deck to play with.

Variations

- Add a spinner to the game. After all of the players have turned over their top cards, one player spins the spinner. The winner of the round is the person who has the card that matches the spinner. (A spinner can be easily made by drawing a circle on the page and using a paper clip for the spinner. Use your own criteria on the spinner.)

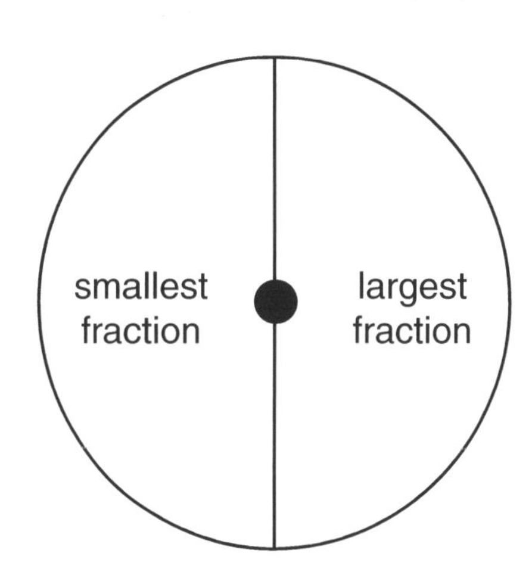

Game 36

Equivalent Fractions Bars

Number of Players: 1 or more

Skill: identifying equivalent fractions

Materials

- crayons or markers
- several sheets of 1" graph paper

Directions

- Write a number at the top of the graph paper. Write the different fractions for that specific number on the left side of the paper. Block off the same number of spaces on the graph paper. (See the example below.) Color the correct number of squares. Write each fraction in its simplest form. (Divide the numerator—top number— and the denominator—the bottom number—by the same number until the fraction cannot be simplified any further.) Write this number on the right side of the paper.
- Make an equivalent-fraction bar page for each of the fractions ranging from 1/2 to 12/12.

Fractions for 10

1/10											
2/10											1/5
3/10											
4/10											2/5
5/10											1/2
6/10											3/5
7/10											
8/10											4/5
9/10											
10/10											1

